The Spiders & Spirits of Petunia Manor

by Yvette A. Schnoeker-Shorb & Terril L. Shorb

NATIVE WEST PRESS

On the cover: Long-Bodied
Cellar Spider *(Pholocus Phalangioides)*
carrying eggs.

**The Spiders and Spirits
of Petunia Manor**

A Native West Press book
December 1996

ISBN 0-965-38491-8
Library of Congress Catalog
Card Number 96-92728

Cover/Title page photograph:
©1996 Phillip Roullard
Design: Amanda Summers
Production: Triad Associates
Printed by SouthPrint International

Manufactured in the
United States of America

Contents

This book is dedicated to both
our beloved human families and
to the extended family of creatures
we have known and loved.

Beginning Life at Petunia Manor

What can be said about 400 square feet of converted shed with gaps under the cheap wooden doors that were wide enough to let in, *and did*, everything from full grown alligator lizards, to Jerusalem crickets (the kind with inch-long, striped abdomens that resemble those of hornets), to night crawlers—except that it was a privilege to have dwelled there.

Fate brought my husband, Terry, and me the blessing of Petunia Manor because we had both chosen, in the spirit of California's uncertain economic situation in 1988, to quit our jobs and go back to complete our college education. Fate had determined that our price for knowledge was poverty. And, because Petunia Manor sheltered us for four years, our living circumstances, including physical poverty, endowed us with knowledge, a type of knowledge almost antithetical to that which was perpetuated by the prestige paradigm of academia in which we found ourselves.

To some, I suppose, the outside area which Petunia Manor shared would have been an embarrassment, a sign to the world that the human inhabitants were stereotypical white trash. The shack-like, little house, with its inside faded, yellowish gray-stained linoleum floors and its outside, weathered white paint peeling, sat on the flatness of no foundation. The tacky box seemed to grow out of a block of cement. The concrete square was surrounded by roughly a half acre of dry dirt, rocks, dog and cat scat, gopher holes, and a lone Redwood. A few blades of grass and some hearty weeds had managed to survive as a lawn—much traveled by creatures the size of which for whom grass blades would be shade trees.

The closest house to Petunia Manor, and for which the manor had originally been built as a shed, was the landlord's house. Located some yards to one side of Petunia Manor, and separated by a narrow alley, this main house shared the lot.

At the end of the back part of the lot, separated from civilized land by a chain fence, was the wild uplands area of a local wetlands. This accounted for the constant air traffic day and night, depending on the species' nocturnal or diurnal inclinations, of ravens, hawks, ducks, herons, and owls over the manor. In addition to serving as a stopover and feeding station in the migratory route of certain species of feathered creatures, the wetlands provided a breeding area for considerable forms of insect life, as well as protective habitat for rodents, reptiles, and amphibians. As we came to find later, most of the human residents of the old neighborhood on which the uplands area bordered fancied the deer and the herons, but the smaller creatures remained of an unspoken status—even though they, particularly the numerous species of spiders, dwelled in abundance.

Because we were both inclined toward fascination with creatures, especially those that crept and crawled, we were not distressed by the fact that the first residents we encountered at Petunia Manor existed in great numbers, mostly on the old broken down porch, and contained rear "pinchers," the females' straight and the males' curved inward. The earwigs went about their lives mostly outside the manor, but occasionally they ventured onto the floor mattress upon which we slept.

What did distress us as we moved what was left of our possessions (as the space could accommodate only one third of what we had owned prior) into Petunia Manor were the bug bombs placed neatly in every room. For days, the little house was perfumed with the sickeningly sweet scent of insecticide. Also, ant stakes, placed like toy army men about to attack, completely surrounded the outside of the house.

The poor, sweet house that was to be our home seemed to have no dignity by social standards. Neither did the little manor have the aesthetic appeal of a picture-perfect, newly painted house surrounded by a white picket fence and colorful flower beds, not even petunias. Our new living quarters, the house that offered shelter, was definitely not quaint. Thus, we decided to verbally enhance the dwelling with the charm of a name: Petunia Manor. And, when we thereafter spoke of our little home, as long as the listeners never laid eyes on the manor, the house was endowed with a picturesque reputation.

For those friends over the years for whom the reality of Petunia Manor was inevitable, that is, those friends who actually visited the house, arachniphilia—or, at least, a neutral feeling toward the eight-leggers—was a mandatory characteristic for mental survival. Once the house had a chance to air out from the bug bombs, which were immediately disposed of, Petunia Manor came alive with one of its best live-in qualities, the spiders.

The Spiders of Petunia Manor

It was the spiders of Petunia Manor in whom I entrusted faith in my psychological well-being. Their smallness and seeming serenity were in contrast to the ever looming, single human presence that moved onto the acre next door only a few months after our first and only *peaceful* year of living at the manor. On the other side of the manor from the side on which the landlord's house was located, and separated only by a worn, wooden fence right outside our front door, the acre and its loud, aggressive occupant became a constant shadow in our lives.

Our attempts to study were now often interrupted for hours on end by roaring, death-themed, canned music from hell in the form of a truck, radio blasting, parked in the middle of this new neighbor's acre. Sometimes the interruptions of study, as well as sleep, came from the whimpering and howling of his dog, whom he had tied to a contraption rigged up to a tree stump on the other side of the fence outside our front door and away from his own trailer. When we complained about noise, a mistake we repeatedly made before we realized with dismay the man's temperament, he responded with less than encouraging words to Terry. After the fourth encounter, we realized that, while there would be a temporary increase in the current noise, the man's responses seemed merely a venting of his anger. During our second year of dwelling side by side on neighboring properties (Terry's and my third year at the manor), our familiarity brought about in the man a slight change. After he felt he knew us, the man's shouted epithets—his natural reaction to being offended by complaints about his noise—were oddly compensated for a few days after they occurred. There would be a knock on the front door at four in the morning.

We were often up at this dark hour because it was the only quiet time that we could read or study or work on term papers. Terry would motion me back as he addressed the visitor—both of us knowing (after the first odd incident of our neighbor's attempt to put things right) who it was. The small, stocky man would be standing with an offering. "I thought you might like some farm fresh vegetables." Sometimes he offered "farm fresh" eggs. Terry was quickly gracious in taking the offerings, although we never ate the food.

At the times of these offerings, the little man seemed oblivious to his surroundings, and I found amusing the fact that a triangle of black widow sisters surrounded our front door. The neighbor, however, didn't seem to notice the large, shiny black widow who lived in a little hole right above the door, and who often webbed out to the center of the porch roof—where she remained active in the night air, directly above the man's head. Nor did he notice the two black widows that occupied respectively areas near the lower corners of the front door, their webs often catching on the man's boots as he stood there. I always fancied the three black widows to be our guards, except that it is the nature of widows to be extremely shy and with tempers that are in no way comparable to the strength of their webs. When threatened, they withdraw.

The cellar spiders, on the other hand, though gentle-natured most of the time, were not shy and could be real warriors when threatened or when hungry. Some people confuse these spiders, because of their long, thin legs, with daddy longlegs (also commonly called harvestmen), but unlike the body of the daddy longlegs, for whom the cephalothorax and abdomen are combined, cellar spiders have separate heads and abdomens.

The cellar spiders had built a network of webbing that connected like walkways every corner of the ceilings of all rooms of the manor. The males, with their long, grayish brown, lanky legs and large, knobbed pedipalps, would use the pathways at night to reach the females. I would often get up during the night to watch the females whom the males were going to visit. The spider ladies would be in their webs weaving, their large, yellowish-bottomed abdomens with the centered, single stripes and pointed, black spinnerets moving rhythmically, often almost in synch with the motion of each other throughout the rooms of the house. It was during one of these nights that I first encountered Goldman and Webman, two females who arrived as youngsters at the same time, and who dwelled in the corners of the bathroom for almost two years. Goldman was named for her color when she was little—a golden ball with thin, brown, black-jointed legs attached. Webman, almost a darker twin of Goldman but missing one leg, was named for her unusually prolific web spinning.

The two had come in through the little bathroom window at sometime during one autumn night. I walked into the bathroom fully expecting to see these two spiders, as I had already seen them enter, walking on the delicate, blue curtains parted along the sides of the window, in a dream from which I had just awaken.

Goldman became a familiar night friend, her sheet web being attached from the window curtain to a point in the middle of the low ceiling to the light above the mirror over the sink. I would often watch her weaving in the wee hours of the morning. I don't know exactly when our relationship began, but if she sensed that I was there, she would stop mid-swing and rush upside down along her web, coming to a complete stop almost at, only slightly above, my face.

After awhile, I saved Goldman time in having to sense my presence by talking to her such that the vibrations from my breath and voice would immediately bring her down to socialize. The sight of Goldman, long, slender legs suddenly stopping mid-weave and body with its abdominal black stripe facing upward, running down her web became a familiar intermission from my dreams. We would look at each other, face to face, her palps pointing upward. I wondered what her eyes (all six dark specks of them) revealed of me. I wondered how she "heard" the vibrations of my voice. Often I would gently touch the tiny, stick-like foot protruding from one of Goldman's front legs with my finger, and she would move another of her long legs here and there, over and about the tip of my finger. The first time that we had this type of contact, I thought that she was going to attempt to wrap my finger for food, but it soon became apparent that she was touching merely out of curiosity, or intuition, or for some other reason unknown to both of our natures.

I soon learned to discern Goldman's temperament, and this through her bouncing ability. Typically, cellar spiders use bouncing as a form of defense, bouncing in their webs so fast that the motion can make them invisible to predators. But Goldman had her own bouncing code for me. A slow bounce seemed to be a friendly communication. She would often bounce in slow motion while sensing my finger with her leg. A medium-paced bounce meant that she did not want to be disturbed. This type of bounce usually occurred when I talked to her, but she did not come down her web to visit. And a frantic bounce indicated that my presence really irritated her—and this was particularly so during the time span that her abdomen was developing into the shiny, bloated state right before she produced her egg sac or, in her youth, a few days before and directly after molting.

Goldman and I had something natural in common: hormones. Mine triggered menses, and Goldman's triggered molting. (Male cellar spiders molt, too, but not as many times as the females, as the females have more life time to accommodate the process.) After having watched this process that Goldman's hormones set in motion, that is, the grueling hour or two of struggling out of her exoskeleton, I suspect that I shouldn't complain too much about the monthly mood swings associated with the minor shedding of my uterine lining.

I could always tell when Goldman or Webman was about to molt, and I had the fascinating (to me) experience of witnessing most of their respective molting periods. Webman, the shyer of the spiders, seemed almost stoic during these times. But Goldman definitely was the one with the "mood change."

A few days before molting, Goldman's appetite became voracious. Her sheet web became littered with the silk-wrapped carcasses of moths, mosquitos, and long, red-winged, flying insects that ate mosquitos. During these times, I never picked the remains of insects out of Goldman's web, as I did—because of their distastefully sad appearance—when she wasn't "pre-molt." Out of respect for her coming ordeal, I waited until the dried, grayish-covered carcasses had been kicked out of her web to the floor before I removed them from the bathroom.

Goldman and I had almost no communication during the days before molting. Goldman spent her time capturing food, biting and wrapping it at night. During the day, she remained attached by her fangs to her food; thus, she continued to feed as she slept. Spiders, with the aid of their venom, digest prey outside their bodies. An automatically occurring pumping motion within the body allows the broken down contents to continue being moved into the spider's stomach.

When the time came right before the actual shedding of Goldman's exoskeleton, I always felt abandoned by my arachnid friend. Goldman would crawl up to the most distant, highest part of her web, with the rear end of her abdomen such that its spinneret faced the world. I knew that next time I saw Goldman's sweet face, with those familiar palps and those two triangles of three eyes apiece, would be after she had molted and had had some time for her body to harden.

While interested in her molting, I always felt sorry that she and Webman had to go through this necessary form of growth. The process needed to allow the succession of exoskeletons that could accommodate the bodies of growing spiders seemed stressful, if not to them, to me. Upside down, Goldman would hang, pulling downward out of her now undersized exoskeleton, the deed sometimes taking hours. Mid-molt she looked so vulnerable with her new, clear, thin legs, only halfway out with the tips still bunched up together in the old, exoskeletal molt while she struggled to be free. I often closed the window at these times, just in case she was as sensitive to wind or particles as she looked.

Once Goldman had completed the task, she usually remained close to her molt for days following. Often, she would keep her molt near her, almost like a security blanket, one or two legs always touching it or in touching distance, even when she ate. Webman treated her molts the same way. I don't know if this behavior is particular to these two spiders or is typical of all cellar spiders.

Spiders, due to their shorter life span, grow up faster than humans. Goldman soon completed her final molt, the top of her long, oval abdomen now forever patterned with symmetrical swirls of dark gray lines. Full-grown, and with her spindly legs providing most of her length, Goldman was roughly two inches long. And at the time of her final molt, I knew that Goldman had become my elder.

She represented to me peace, normalcy, serenity, and wisdom. And I was thankful for her presence many times during our stay at Petunia Manor, particularly those times when neighbor problems became almost unbearable, such as the day of the lamb incident. The neighbor had bought a full grown market lamb and, in the spirit of self-sufficiency, fed and eventually slaughtered the beast himself—directly behind the fence facing our front door. I had the unfortunate luck of having to be home at the time to work on a term paper, thus was surprised by and exposed to the prolonged ruckus, loud cursing, and various other noises that indicated our neighbor may not have been terribly experienced at this old pioneer task. I could hear him talking to himself, but I was afraid to go outside to peek between the boards of the fence. Instead, I sat quietly on the white throw rug on the bathroom floor and watched Goldman quietly sleeping in her web.

More routinely, the neighbor seemed to relax by driving about his yard on a backhoe, digging up heaven knows what. Trying to study to the rumbling background of the backhoe was a common frustration. I often sought the symbol of quietude in the sight of the cellar spiders hanging gently and *sanely* in their webs.

During one of these times, I came upon Goldman, who had "been with egg sac" (held protectively by her fangs) for two weeks. The shell of the white sac was cast to the side of her. She was hanging peacefully in her web below a universe of little spiderlings, newly detached from their egg sac and loosely centered in the strands of silk directly above their mother. Goldman, sensing my presence, began a slow, rhythmic bouncing. Her sweet spiderlings, all legs with clear bodies barely visible, were so tiny and yet had so much presence. The thought occurred to me that the opposite could be, as well. Something, like neighbor problems, seeming so big, could be made small. And at that moment I realized that, in my scheme of things, Goldman and her babies were a much larger, important presence in my life, and now in my memory, than the human distraction next door.

And, thus, through Goldman and her young, I learned, not only to deeply appreciate little things, but to know that sometimes it is wise to make—at least in the mind—big, bad things small.

Spiderlings and Toothbrushes

The third spring that we were at Petunia Manor, we received a cheerfully presented but devastating note from the landlord, who had a strong sense of responsibility and an even stronger fear of any life form smaller than a cat, that a professional exterminator was coming to spray his house and *ours* for spiders. Because the landlord, a conservative, middle-aged man, had considered us in all other respects associated with "good renters" a nice, normal, dependable couple, Terry and I had purposely kept the man unaware of our fondness for little, flying, creeping, and crawling things. Upon seeing us leaving for the university the morning after the arrival of his note through our mail slot, he yelled over in a neighborly manner, "Mornin'. Did ya get my note?" I could see by his friendly smile that he had anticipated that we would be thrilled by his news. Evidently, he had come across a black widow in his garage and had also heard from a neighbor down the street that brown recluses could *possibly* be in this part of California. The landlord did not have a key to Petunia Manor, and he felt that one of us should be there when the pest person arrived. He was apologetic to us because he could not find anyone to spray until the following week. Both Webman and Goldman were with egg sacs.

While I had read in different books that cellar spiders, particularly the species of Goldman and Webman—Pholcus phalangioides—had only one egg sac during a lifetime, I observed that, over a period of a year and a half, Goldman had two egg sacs and Webman three. The most memorable group of spiderlings manifested that spring when the two lady spiders developed little ones during the same period. One huge sheet web, which began with Goldman's original web and expanded outward beyond the center of the ceiling and almost to the bathtub curtain directly opposite the window, contained both spiders. Each, egg sac attached protectively at the mouth, occupied her respective corner.

The beginning of the egg sac has always been a mystery to me. I only once observed the beginning stage of the thing that would develop into spider babies. One night, walking in on Webman, I saw a brownish, mushy substance that she was wrapping carefully. The grainy substance did not look like prey, and the mass was taking on a spherical form and whitish color as Webman wrapped. Because Webman's abdomen, which had been huge for weeks, was now shriveled such that the abdominal stripe was not visible, I knew that the mush, though not appearing like a group of individual eggs at all, was the beginning of her egg sac. I never saw from which part of the body the mush came. Usually I first observed the sacs, a few hours after they had been produced, when the individual egg compartments of the sphere were noticeable. From then on, the most extreme change in the sac, besides that it grew larger, was the increasing degree of "hairiness" that was actually the formation of hundreds of thin, little legs forming as the spiderlings matured. Right before detachment, the sac became a ball of little beige, hair-like legs moving and brushing up against each other. I wondered if each little pinpoint of consciousness was somehow aware when one of its legs accidentally touched within the tangle on its place on the sac another pinpoint's leg.

That time at which both spiders were with spiderlings, the mothers had produced their sacs within the same evening, but the two groups of young had developed at slightly different rates, Goldman's detaching from their egg sac first. A few days later, Webman was also centered under a little universe of starlike spiderlings. After another few days had passed, the very first molts remained clustered where the two little universes of spidery stars had hovered, and both groups of small, hairy dots began to wander around the entire web. I could not discern which spiderlings were Goldman's and which belonged to Webman.

The baby spiders explored with purpose their inherited web, the tiny, leggy things bobbing up and down as they walked. Eventually, the spiderlings began to experiment with webbing. They would move along the web, occasionally stopping to push the bottom of their abdomens against a strand of mother-silk upon which they had been moving. After the attaching motion, they would rise up, seeming to almost strain to pull away from the web, and walk away, continuing to explore.

It was during this time of spiderling exploration that I woke up one morning to find that one of my current professors, who found amusing and worthy of chitchat my interest in spiders, was wrong in his assumption that human saliva is deadly to spiders. I switched on the bathroom light to find that our white toothbrush holder above the bathroom sink, particularly the bristles of the two protruding brushes, had become a point of attachment for a shared, silken highway with many lanes leading up to the main web. For some curious reason, those babies that did not disperse to other rooms of the house had found appropriate the formation of a triangular sheet web that connected the tips of our toothbrushes to the bathroom mirror and to the mother web.

Staring in delighted fascination, I watched as the little creatures spun and dropped and pulled and walked around as if they had owned this world forever. Each spiderling, whose dotlike, little cephalothorax could now be seen without straining my eyes, seemed unaware that he or she was contributing to the same creation.

Their mothers were not as fascinated as I. Goldmam and Webman, who were up in newly spun, separate webs and facing, respectively, different corners of the bathroom ceiling, had both captured prey at some time during the night or early morning. Both were quietly attached by their fangs to their wrapped, dead captives. Aside from an occasional, slight shift in feeding position, both spiders, who typically ate once (if at all) during the weeks of holding their egg sacs, were motionless. I wondered if Goldman and Webman were dreaming as their shriveled, tired bodies, through the unconsciously automatic sucking and pumping motion that nature gave them, took in much needed nutrition.

The youngsters, on the other hand, were anything but motionless. A few of the spiderlings had found the light blue drinking cup next to the toothbrush holder, and they began to include the rim of the cup in their creation. I quickly ushered the determined little arachnids off the cup with the tip of my finger, placing it in their various paths. Upon sensing my finger, some of the little ones quickly scurried up to the safety of the established webbing on the toothbrushes and the mirror. But more than one of the spiderlings attempted to web my fingernail. And one tiny spiderling suspiciously pushed the pinpoint that was his or her head—fangs (though too small to see) first—against my finger. With another finger, I carefully nudged the little beast up onto the front surface of my fingertip and placed the spiderling up onto the mother web.

About this situation, there was only one thing to do. That evening the drinking cup sat next to our new, additional toothbrush holder, complete with two new toothbrushes, on the opposite side of the sink from the original holder. But this new holder and set of brushes were to be ours, as the spiderlings had their own—and many a growing spider got much use out of this oddly web-attracting territory over the year.

After this incident, while I was happy to learn that human saliva did not hurt cellar spiderlings, I was even more distressed than before at the prospect of the pesticide spraying for which Petunia Manor was fated. A week had passed since last speaking with our landlord and the fatal date had been decided. After desperate hints to no avail that I had horrible reactions to certain chemicals—insecticides, in particular, I had offered to be home to let in the pest controller. The landlord attempted to firmly brush aside my worries by saying that, these days, the sprays are "so safe for humans that you could drink them." Pleading with the landlord would have definitely aroused suspicion. He had no idea of the company that we kept when inside his rental. And we had continued to hide well this secret. Upon hearing the landlord approach, or whenever we found him knocking at the front door, we would quickly step out onto the front porch and closely face him in a friendly manner that caused him to unconsciously back up to allow for the comfortable body distance for socializing. More importantly, placing our presences in front of him protected the inside of Petunia Manor—and the dozens of cellar spiders up in the ceiling corners—from scrutiny and subsequent consequences *for all of us*.

However, there was no way to hide our arachnid friends and their thick webs throughout the house from the pest control guy. He was being paid to do a job, and we weren't the ones paying him.

On the dreaded morning the green truck, on the side of which was painted a giant, black spider with a red slash through it, arrived. The young, blond man, who wore a green jumpsuit that advertized his pest control company and carried a large canister and a box with various nozzles, got out. He was lead by the landlord into the main house. Fifty minutes later, both men were outside my front door. Upon hearing them approach, I quickly stepped out to greet them, telling the landlord, as I ushered the pest control guy into the house, that I would send the exterminator back over to his house after our rental had been sprayed. The landlord seemed pleased by my cooperation and wandered off, obviously having other tasks to tend to that morning.

Once inside Petunia Manor, and after giving the corners of the living room a quick glance and shaking his head, the man went directly to the kitchen and began to study the areas along the ceiling, floor, and various corners. Without looking at me, the short, somewhat stout young man shook his head again, saying with a sigh, "It's a good thing we're doing this. You sure have a lot of spiders in here. Should have mentioned something about the problem to your landlord earlier." He poked his head behind the refrigerator and stove, then looked over the outside of the cupboards. "Any problems in your cupboards? Get a lot of spiders or ants in there?" He was crouched down looking in the lower cupboards, but he turned his head so that he faced me directly, his light blue eyes emphasizing his questions, which were really one question. "How about the other rooms of the house? Bedroom? Bathroom?"

I quickly changed the subject, inquiring as pleasantly as I dared, "Hey, what do you have against spiders?"

Taken by surprise, the man raised up. "Nothing," he smiled. "I love spiders. They afford me a good way to make a living." I suspect that he was attempting to be humorous.

"Well," I smiled back, "I'm quite fond of spiders, too—living spiders, that is."

The man was definitely amused. "You are joking, Lady, right?"

"Absolutely not." I grabbed three twenty-dollar bills that had been set neatly on the kitchen table that was, because of its large size, in the living room. "Our landlord never comes in here to visit, probably because, as you can see, the space is so small. He's a nice guy—respects our privacy. Doesn't even have a key." The pest guy was staring, more out of curiosity than anything else, at the bills in my hand. I continued, "How about if we just *say* you sprayed? I'll tell him that you did a great job, you'll get paid from him for spraying the spiders ... and paid from me for letting them live." I offered him the sixty bucks, adding, "And I'll throw in a cup of coffee. This is a small place, but you should probably be here for about ten minutes at least to make it seem like you did something."

After shaking his head back and forth slowly in amused bewilderment, the man smiled. "You don't have to pay me, Lady. Just tell him I did a real good job." I could tell that I had presented him with a dilemma, and that he was extremely uncomfortable with his decision. Evidently, though, he had a kind heart.

Two days later, the landlord called. "I'm gonna have to have that young man back. I think he missed some areas in my garage. Spiders are already crawlin' 'round in there. Saw one in my house, too. We should get 'em second spray for good, though. How's your house lookin'? Need the guy to do some touch up sprayin'?"

The tone of my voice in responding to my landlord held nothing but surprise. "*Really?* Well, that pest guy must have done a great job on our house." I watched as a "teenage" cellar spider bounce-walked across the phone cord and up the wall. "There's not a spider in sight."

The Three Sisters

She arrived one autumn night when Petunia Manor lay shrouded in tule fog. A big sugar moon shone down on the Laguna. I got a drink from the kitchen tap and peered out the window. It was as if the house was packed for shipping in a layer of cotton. I turned away from the window and felt something brush my toe. Soft, supple, dry—like strands of long hair. There was a tearing sound, like sheets of parting Velcro.

I switched on a light and beheld a sight that, thirty years before, would have given me a yelping start. There, down between the base of the cupboard and the frame of the door, busily inspecting the damage wreaked by the intruding toe, was a mature Black Widow spider. The remaining filaments of webbing glistened much like moonlight on the fog outside. I crouched down beside the web and watched as the spider moved deftly, a silvery line of new silk emerging from her spinneret. Like a living shuttlecock she spun and wove, mending her torn sheet web, her house of silk.

I watched the spider's graceful movements. So purposeful. So...innocent of the huge presence which observed her. She was merely a future mother working tirelessly to create a place for her expected children. You've come a long way, I noted to myself, from the days when the mere sight of one of these obsidian ladies would've sent me scurrying.

In thinking of her innocence of human perceptions of her, I was reminded of our own naiveté as children, and what a strange and beautiful power is generated by that kind of magical consciousness. Imagination holds a transformative power that the old alchemists never found. Consequently, the kiddie cosmos is inhabited by a bestiary of outsize creatures and invisible (to other, less sensitive eyes) presences that are welcomed by children as playmates. No xenophobia there! But then of course come the stories, the indoctrination from the adult world. The cautions and prohibitions and fears passed on down from one generation to the next. Soon, the cotton candy landscape of the child's imagination is haunted by fearsome creatures who harbor razor-sharp teeth and a taste for young human flesh.

Among the denizens of the darker region below the houses of childhood one could expect to find the feared Black Widow. Most of us kids knew that next to the Jolly Roger pirate flag, the crimson hourglass worn by Mrs. B. Widow was a symbol of sure death. One bite and you're history. Kids passed from one to another stories about kids getting bitten, swelling up purple and rotund as plums, and then exploding in a shower of gore. There were whispered tales of Black Widows in old Mr. Miner's garage up the street that were the size of garbage can lids. These were the things we believed about the shiny, black ladies that hung in their rude cotton lairs in the darkest corners of our childhood days. Never mind that I never met a victim or survivor of a Black Widow's bite, or attended one's funeral. In fact, I was 24 years old before I met another human being who had actually received the "fatal" bite. So astounded was I by the living presence of this person—I married her! Well, actually, it was for many other wonderful qualities, but I was intrigued to at last make the acquaintance of someone who had encountered the legendary leggy lady and lived to tell about it.

"My arm swelled up," Eve explained of the bite. "About like it does when I get bitten by a mosquito." I confirmed that the Black Widow's bite is seldom life-threatening, though humans frequently experience painful muscle spasms or sweats or other unpleasant symptoms of the neurotoxin at work. Mortality statistics show, however, that deaths from Black Widow venom are rare. Far more humans in America die each year from falling down their own stairways at home.

Not all Black Widows are widows. The legend is that they often consume their mates after consummation. But in the years that Black Widows shared Petunia Manor with us, we never saw the Misters who romanced these notorious spider women. Nor did we ever find a suitor's shriveled remains beneath a Widow's web. Somehow, though, the courtships occurred. Because one day upon our return from a month-long visit to Wyoming, we beheld in the corner web of Mrs. Widow the presence of new life. For there upon the stout strands of the nursery were a few dozen or so baby Black Widows. Their juvenile attire was very different from the simple, elegant ensemble of their Mom's. These youngsters were gaily clad in harlequin hues, with bright bands of coral pink and sand dune gold painted in elaborate swirls upon their rotund flanks.

Well. We held a conference and aired out concerns. Though we welcomed all spiders to Petunia Manor, there was the purely practical issue of how difficult it might become to avoid accidental contact with Black Widows, particularly once all these kids had grown up and set up websites of their own in other recesses of the Manor. We decided to relocate the youngsters to safer quarters—for them as well as us.

We then talked to the mother about her children and told her why we were evicting her offspring. We begged her forgiveness. One never knows in such moments if the communication is received, but we have long felt that it is right and proper to tell other beings why we are disrupting their lives. It is mere common courtesy among companion creatures who happen to share a habitat.

The relocation began, and I used a popsicle stick to roll up strands of the nursery web, along with a spider kid or two. Some were then unrolled and draped like brightly colored holiday ornaments along the sagging, wooden fence on Petunia Manor's eastern perimeter, right outside the front door. Others of the relocatees were released on the wooden front porch—an open air and peeling affair that afforded all sorts of little nooks and crannies. From there, the kids could choose to stay or to move off to other lands.

When the last one had been relocated, I stood watching it move gingerly over its new surroundings. It soon disappeared into a beige-painted rift valley between two clapboards on the side of the house. Months passed. We became absorbed with school and with a looming danger that made Black Widows seem harmless by comparison.

The northern stretch of Petunia Manor's half-acre huddled on the shoulders of a vast, freshwater wetland. One of the few such habitats of that size in all of Northern California—most of the others having been long since drained and developed. The richness of the aquatic and mammalian life on the uplands made the Laguna an ideal habitat for early-day humans. Descendants of what was once the largest village of Pre-Columbian peoples still dwell in the area. They say that their babies are buried in the pale soil, beneath the wild flowers and comings and goings of foxes, raccoons and other inhabitants of the gentle slopes that rise above the dark green marshes.

The complexion of the Laguna through the seasons was as complex as its ecosystem. Summer baked the surrounding foothills to a nutty brown while the moist lowlands stayed a luxuriant green, dappled by wild blackberry thickets and stately oaks. Spring brought to the Laguna a riot of growth, including blooms of rare flowers and vernal pools in which blossoms of delicate gold bloomed close to the seeping ground. During the early evening hours the air was festooned with the love songs of thousands of frogs—a Milky Way of music whose aural intensity overcame the thrum of traffic from the nearby state highway. Winter rains transformed the Laguna into a shallow lake peppered with thousands of preening waterfowl. Herons and egrets patrolled the edges of the water, and land animals used the uplands for passage around the rippling new prairie of water. Many was the evening we awoke to the sounds of footfalls. In the misty morning we would find tracks of deer, rabbits, foxes, and others who make their living under the sheltering stars. It was to this precious crescent of land—locked between the lapping waters and the concrete and steel fences of the urban sphere—that the threat advanced. A developer saw the living resource not as home for four-legged and feathered creatures, but for well-heeled commuters who would pay dearly to live in massive houses at the edge of one of the last freshwater lagunas in the region. Suddenly, the very sustaining presence of nature's last stand in this increasingly urban region was in jeopardy.

We despaired even a fight, knowing that our school and employment already consumed our available energies. Not to mention it was said that the developer had a lock on the proposal because it was situated in the city's residential development zone. A done deal. We did manage to drag ourselves to early scoping sessions, heard from the councilmen favorable to the project that it should be approved quickly to allow the developer to get in there with his dozers before the spring rains.

We walked home one evening from the council chambers on leaden feet. We walked past Petunia Manor to the end of the acre on which the dwelling sat. We gazed out over the great expanse of blackness that was the Laguna. Above us, their primeval fire not dampened by urban lights, stars dropped diamond kisses upon the slumbering grasses and reeds, and upon the innocent creatures huddled in innocent slumber. We made a promise that night that breathed a prayer out across the Laguna: somehow, we would help.

We walked slowly back to the dim beacon of the lightbulb which glowed over Petunia Manor's humble front door. I fumbled for the key. Eve tapped me on the shoulder.

"Look!"

There, squarely above the door, suspended in the amber light like a dark jewel, was a Black Widow spider.

"Over there!" Eve pointed to the bottom left side of the door. Just above the crumbling cement pad and the rotting base of the lowest sideboard was another Black Widow, also suspended in her web. As if pulled by a strand of webbing, our gaze was intuitively pulled to the opposite side of the door frame. Amazingly, suspended on yet another web between the house and the leg of an old table that we occasionally used for porch meals, was a third Black Widow. The portal to Petunia Manor was bounded by a living pyramid of arachnids. Could this be happenstance?

The thought came to us simultaneously. These ladies were here to help, to protect the Laguna and to watch over us as we did the same. It was clear as starlight to us both that these Three Sisters had appeared as a sign to us that the impossible would be made possible. The Laguna now had powerful allies in its defense. We were filled with a strength and determination and hope that the Laguna—now the cradle of wild kingdoms hereabouts—would be preserved. We went to bed with crescents of smiles on our faces. In a dream the Three Sisters traveled out across the tule fog hovering over the Laguna, weaving as they went. They transformed the fog into a magical blanket of protection.

I awoke to strong sunlight filtering from the redwood tree just outside.

"The Laguna will be saved!" I said it matter-of-factly, having no rational reason to believe it nor any idea of what the plan of action would be. That was in 1990. The story of the events that transpired across the intervening years and surrounding the battle for the Laguna is one that will require a book of its own. Suffice it to say here that the inspiration of the Three Sisters and Goldman and Webman and Magic and the scrub jays and the Redwoods and the distant ravens and the other creatures who flew, crawled, and webbed into our lives sustained us through an interminable thicket of meetings, hearings, neighborhood organizing, campaigns for public officials sympathetic to the Laguna, research, letter-writing flurries, and long, quiet evenings of near despair. All the while the Three Sisters hung in their jeweled webs by Petunia Manor's front door, weaving, weaving, weaving.

Fate intervened to take us away from Petunia Manor in the autumn of 1992, but the battle to save the Laguna raged on. We kept in close touch with our human friends who bravely carried on, read the news clippings they sent of the ups and downs and all-arounds of the conflict. Then, one day, nearly three years after our departure from Petunia Manor, and five years after the battle for the Laguna had begun, we received a phone call. It was one of our best friends and fellow human residents of that Laguna uplands area. He had become a fearless and ingenious leader in the fight.

"Good news!" he shouted over the phone. A deal had been signed with a local land trust. The uplands of the Laguna were now in safe hands. The conservation agreement said in unequivocal terms that the Laguna could never be developed.

"Long live the Laguna!" I shouted back over the long miles.

That night, Eve and I made a silent prayer of gratitude to all of our human friends and acquaintances who had carried on to victory. And we breathed grateful sighs into the cool, high desert air of our new home and began to say the names of the many creatures in and around Petunia Manor who, we knew, had helped to preserve that beloved place. We uttered special thanks to the Three Sisters who had framed our door with their webs of silken potency. We also thought of something we had recently learned in our new land, from our Hopi neighbors to the North. These indigenous people of the high mesas of Northern Arizona say that they came into the world through the help of Grandmother Spider, who literally wove the world into being. And, we realized, is still weaving.

Land of the Redwood Empire

Just beyond the front door of Petunia Manor lived a member of an empire that once stretched from Canada to the central California coast. This empire has endured for a thousand, thousand generations and has survived every enemy sent against it.

Until now. What fire, flood, pestilence, and drought could not conquer in a hundred million years, a mere two generations of Europeans nearly extinguished.

Our lone Petunia Manor redwood tree, a Sequoia sempervirens, was a magnificent individual. An adolescent of perhaps sixty years of age, the tree was eighty feet high and stood rooted to the earth with a bole two feet across. It wore a bark of shaggy fur the color of old pennies. Her canopy shape was a perfect, slender cone, like one of those Princess hats of Medieval times. A splendid member of her proud race, she nevertheless stood alone, her ancestors having been felled to make way for the little suburb at the edge of the Laguna that included Petunia Manor. This lone Redwood was for us a sad symbol of the plight of her race: nearly ninety-five percent of the original coastal redwood forest has been sacrificed to the demands of humans.

Each morning at Petunia Manor that was not obscured by fog brought a touch of the sun's rays to our pillows beneath the east-facing window of our bedroom. The tree stood outside this wall and its shimmering needles massaged the sunlight, shaped and worked it until it fell upon our cheeks in an otherworldly hue. This wet, emerald light possessed alchemical properties that could pull us from sleep softly and irresistibly as a mother reaching for her newborn. We decided that there was no better way to be summoned from slumber than to have a long, green tongue of dinosaur light slurping up one's forelocks.

Our first encounter with redwood trees was in 1982, on the first evening of our first day in the Northern California region that we were soon to call home. We were elated and numbed by the stark reality of our situation: we were strangers in a strange land with absolutely no contacts, identified prospects, or specific plan of how we were to "make a living" here. We had but a few thousand dollars wrested from savings that would be used up in a couple of months. We were in dire need of a plan. A place to think things through. The busy strip where our motel was located was incessantly noisy from passing highway traffic. I inquired at the office about a good place to conduct a contemplative walk.

I imagined that the proprietor would direct us to the ocean—just a few miles to the West. And indeed he did mention the Pacific coast in the even voice of a chamber type doing his duty to inform the visitor of the best sites and sights. Also, he said, there are the handsome interior valleys where the foothills are festooned with vineyards.

"Or..." And his eyes suddenly widened. "You might want to take a stroll in the land of the giants."

Perhaps it was because his motel was named for the big trees. A pair of juvenile redwood trees flanked the office entrance. At any rate he gave us directions to make our way down the Russian River to the last grove of intact Redwoods in the County. We made a mental note to go there as soon as we put out a few feelers related to immediate survival. In the days that followed we made dozens of phone calls from the steel phone booth outside the motel, we scoured help wanted ads, we filled out applications at the State jobs office, and we knocked on doors of businesses that bore any faint resemblance to aspects of our experience or training. We made copies of resumes and practically leafleted the region. The manager had graciously agreed to let us use his office number for return calls, as there were no phones in the room (in the interest of economy room rates). We listened to the roar of traffic, paced, played cribbage, ate deli sandwiches, and checked the manager's office about fifty times an hour. No--there had been no callbacks. It was then he reminded us of the great trees huddled in their misty canyon.

"It'll make you feel better to walk amongst them," he offered.

The next morning, in early December, dawned crisp and bright blue. The ever present marine layer of fog had burned off early. So we struck out in our little, red Fiat in search of the last elders of the true redwood empire. We were directed up a winding, narrow road from the river town of Guerneville. It was once known as "Stumptown," for the many chest-high remnants of mighty redwood trees felled in the turn-of-the-century Redwood lumbering frenzy of that time.

We rounded a last curve and found ourselves at a deserted toll booth. A sign indicated we could either pay and drive through or park in a nearby lot and walk through. We decided instantly that the only way to experience the big trees was on our own humble feet. We parked and set out. Our footfalls were swallowed up by the looming wall of foliage which towered just beyond the wooden sign to "Armstrong Grove Redwoods Reserve."

We crept passed the first old-growth trees which were as thick at the base as grain silos. They were as columns of some gigantic edifice—a temple to the gods. I was so overcome by the power of their presence I doffed my hat as I would upon entering any sacred site.

We walked in deeper and the blue day was transformed into a strange, shimmery green filtered through two hundred feet of canopy. The soft glow was animated by breezes far up and hidden from us by the canopy. Strong sunlight, we could imagine, was caught, caressed by needled branches and then passed down, limb to limb, until, here and there on the shadowed forest floor, there fell to ground a gilded dollop of light. The effect was like walking on the bottom of a lake. The air shimmied past us in green-blue waves, soft, alive. The quiet air took on the quality of living tissue.

We walked on, more slowly with each step, past ruddy boles whose size told of lives measured in millennia. Some of these trees had been witness to a world before the fall of Rome. And yet, for all their superlative dimensions—more than three hundred feet in height and girths equivalent to a one-bedroom house—we discovered that these ancient ones propagated in a surprisingly humble way.

I had imagined that we would find seed cones the size of basketballs, suspended hundreds of feet up. They would fall with a shriek of breaking branches and thud into the ground like bombs. I idly wondered how many creatures had been crushed by one of the plummeting missiles. And then we came across our first redwood cone. It was right in our path. It was still attached to a sprig of slender, needled branchlets. The color was nut-brown. It was an intricately sculpted seed case. The size of my thumb! Later we were to learn that these gentle giants preferred clones to cones as their chief means of making progeny.

The canyon began to skinny up and the trees got even bigger. Our own voices grew more hushed, smaller, like the secret whispers of mice. Perhaps it was the effect of being shrunk to the Lilliputian scale, or the eeriness of walking among creatures who are more than a thousand years old, but I began to experience an altered state. Things flitted in the gauzy corners of my peripheral vision. Scampering echoed from behind rumpled forms on the forest floor, these appearing to be long-fallen trees or branches now blanketed by lichens and bouquets of ferns. We spied Redwoods whose boles had been hollowed out by licking flames of wildfires. Hungry flames had penetrated the heartwood, but had left the life-bearing cortical layers intact. These cavities were often large enough to admit creatures even as large as ourselves. We crawled into one such living cave, the odor of carbonized heartwood still strong. It was like huddling inside a boulder of coal. We peered out through the black-skinned cleft at the green shapes beyond. A delicious shiver came over me, as I once experienced in childhood. It was the pleasure of being secreted *inside* another living creature—half hidden from the outside world. At that moment, I would not have been the least bit surprised to see a unicorn trot by. Or the tiny footfalls of a Hobbit off to his magical chores. All of the rules of reality seemed suspended. Magic hung thickly in the air. We slowly inhaled ancient ethers and let gossamer thoughts glide through our brains.

We clasped hands and embraced the same thought, simultaneously. If the rules of reality are tenuous, then *anything* is possible! It was an affirmation that grew up around our beating hearts as surely and vigorously as the daughter sprigs growing from the old body (reality) of the parent tree.

"Magic lives here!" my soulmate whispered. I nodded, too giddy to even speak. And then from across the emerald air a sound. A riffle of laughter? Not human. But creaturely. We weren't to guess until years later who the messenger was, but the message itself found its way straight into our hearts: Take care of these trees and all of the trees who dwell in this realm and all of the other creatures who dwell with them, and they will take care of you.

We crawled back out of the tree, still holding hands, and made our way back through the enchanted forest. Carrying with us, for the first time in a long time, the enchantment of hope.

P.S. Within days we both found employment. We were soon as happily rooted in the Land of the Redwood Empire as were the big trees themselves.

Banana Slug

It was as if Zeus had hurled one of his thunderbolts to Earth and it had shattered into shards of pure, living gold. One now lay before us in the ferny path of the Redwood grove a few miles West of Petunia Manor. It was about five inches long, thick as my thumb. It was tapered to a point at one end. This last feature drove into my head the improbable image of the golden spike that was driven into the railhead at Promontory Point, Utah to connect up the cross-stitched ends of the first Transcontinental railroad.

We stood riveted, staring at this shimmer of metallic beauty so incongruous among the soft green fronds of dinosaur ferns on the pink-needled forest floor. And then the bar of gold exhibited an even more remarkable quality. It moved. Well, actually, something on it moved. From the rounded end there grew a pair of projections slender as amber toothpicks. Atop each was a black dot like a flake of pepper. It was as if this submarine-shaped thing had extended two periscopes. And, truly, one now angled in our direction, as if regarding the pair of huge creatures that hovered over it.

It was then we figured it out. We were being scrutinized by an eye. Atop an eye stalk. Atop a Banana Slug. This splendid creature was the most exquisite mollusk this side of the emerald sea, home to the more familiar abalone.

We bent closer, unwittingly penetrating the slug's secure airspace. She quickly retracted her eye stalks and then her whole body retreated into itself until she was half her original length. We moved back a few feet and waited. After a bit one eye stalk cautiously rose up from the smooth, gold head—giving us a flashback to the old TV series, My Favorite Martian. Now the slug extended both eye stalks. Then her body telescoped out and she began to move. Never in all our days of creature watching had we encountered such grace in motion. The slug glided over the prickly texture of the forest floor as if she rode a carpet of air. In her wake was a silvery legacy of slime. It reflected a full spectrum of colors. If diamonds could walk, this would be their footprint. It was a ribbon of rainbow.

All of us creatures of the Earth struggle against gravity to effect locomotion. Most of us simply exert raw muscle power to hurl ourselves at the invisible harness wires of centripetal force. The Banana Slug, however, is a more sublime traveler. Her self-created silk-slime road dampens drag, hushes friction to a moist whisper. Her contact with the ground is a continuous, wet kiss, a cooling caress. She moves with such quietude and finesse that one imagines gravity not even being awakened to its task of resisting the movements of all living things.

We instantly fell in love with this fruitless banana and lay beside her at the great, ruddy feet of Redwoods and watched as she grazed on miner's lettuce. Our eyes were not strong enough to discern her 2,000 or so teeth at work. We were later to learn that Banana Slugs like apples, so upon subsequent visits we brought slices of apple to the banana. The slug would virtually embrace the wedge of fruit with her limber body, and while she ate, it looked like an apple had been embedded in a circle of gold. Once, while we strolled the redwood forest, which grows on the campus of the University of California at Santa Cruz, we happened upon a banana slug grazing just off the footpath. We crouched down near her and spoke to her in soft, friendly tones. A student came along, got off his bicycle to see what we were looking at.

"Your school mascot," I whispered, having seen the pennants hanging in the University bookstore. The young man, who was tall and muscular—athletic looking, bent for a closer look. His eyes widened. Then he unconsciously backed up a half-step. "My God! I didn't know they got THAT big!"

"Yes," said Eve, straight-faced. "And this is just a teenager!"

Months later a state legislator from Santa Cruz county seriously introduced legislation to have the Banana Slug succeed the abalone as California's state mollusk. He cited its rarity, its uniqueness, its accessibility (people can actually happen upon them and observe them, as opposed to the abalone who is seen in its habitat only by abalone divers armed with pry bars). But the abalone forces held off the golden pretender to the throne, citing the abalone's value as a commercial tyro in the state's marine fishing industry. So of course, in the good old U.S. of A., where the race almost always goes to that which can be converted to cash, the slug lost out.

In our hearts, however, the Banana Slug is eminently worthy of distinction. Her example is an inspiring one. For it matters not to her how rough the road of life may be. She carries within herself the means to create her own way, to glide right over that which would otherwise bring her to a halt. And she does this in a way that is friendly to the Earth. What we learned during those years in Petunia Manor, was that, like the Banana Slug, it is possible to go forth purposefully, slowly, peacefully. It is possible to proceed in such a way that one finds what is necessary to life, at no net expense to the Earth over which one travels. That is the best—when one makes one's way in life and leaves nothing but a rainbow in one's wake.

Continuing Life at Petunia Manor

One of my biggest fears during our second to the last summer at Petunia Manor was that I would walk into the bathroom and find either Goldman or Webman, both now very old ladies with many children to carry on their lines, hanging lifeless in her web. This is how cellar spiders who dwell in houses and garages and cellars die. Because it is more common to find males, who don't live as long, I had over the years many a time come across the still, dried bodies—small abdomens quite emaciated and almost gone—with legs curled stiffly inward. I would remove the dead spiders, as I removed abandoned molts, from the web and put them outside where they could be picked up by the wind.

But I never found Webman nor Goldman dead. One weekend morning near the end of summer, I awoke to the typical sound of three scrub jays who, in addition to announcing with their shrill voices the passing of the sunrise, had taken to pecking on the low roof directly above the bedroom of the manor. I walked into the bathroom and noticed with alarm that neither Goldman nor Webman was in her web. The two spiders who had dwelled with us for almost two years were not among the spiders in the bathroom. Both spiders had disappeared together as curiously as they had come to us. Although I left their webs up, I collected the part of Goldman's web that formed a sheet across the top of the mirror. This I bunched up into a small, silken clump and kept as a reminder of her.

I wish that I had kept something from all the creatures—spider molts, discarded exoskeletons, feathers, shed lizard skins—that touched our lives at the manor. While to the landlord, had he known who dwelled with us in his rental, Petunia Manor might have been a menagerie of living things feared, to Terry and me, perhaps because the essences and occurrences of life within were a road away from the rational, the manor bordered on the realm of magic. And *Magic* was the name we gave to the little, hibernating guest that insisted on sleeping under the kitchen throw rug during our last winter at the manor. Magic was the second alligator lizard—that we know of—to find its way inside Petunia Manor. The first lizard appeared, oddly, earlier the same day that Magic appeared for the first time one summer.

There are three ways in which a northern alligator lizard, with those penetrating, catlike eyes, one on each side of its pointed snout, looks at a human. The long, lengthy-tailed, low-to-the-ground, brown, gray, and black checkered lizard may sense the human. The lizard might then, running swiftly on its little legs with the motion of a fish, stop occasionally and turn its head to glance back with one eye or the other before disappearing into safety. Or, the lizard may freeze and stare quietly and intently at nothing, appearing not to see the human—but the lizard knows the human is there and, if one looks at the lizard closely, one will detect a slight change in the angle of the eye nearest the human. The third way a lizard looks at a human could have been demonstrated by the way a large, slender alligator lizard watched me from the center of our white, bathroom throw rug, where she sat calmly motionless one morning when I was about to step out from the shower. She had not been there moments before when I had stepped in, and her presence took me completely by surprise.

Head tilted to one side, the lizard stared up at me with one eye, her gaze following my foot, which I quickly withdrew back into the tub for fear of frightening her. The lizard then tilted her head to the other side, eye focused on my face and looking me specifically in the eyes. Her body remained motionless, except for her head, which tilted from one side then to the other, such that either eye continually watched my movements. On occasion, her thin tongue stretched out from her slightly open mouth, the tips of the fork at the end sensing particles in the air before withdrawing back into her mouth. I have heard that the pendulum motion of the head and the rhythmic swaying side to side of some snakes and lizards as they glide or walk is caused by a sensory organ in the upper palette of the mouth. The tips of the split tongue collect information from air particles and, after the tongue brings the particles back into the mouth to be sensed, the side of the fork which has encountered the most particles is the side to which the reptilian head will tend to sway the next moment. But the only head movements by the alligator lizard on the rug were those that allowed her eyes to assess my immediate activities.

Delighted to see such an unexpected gift, I moved slowly. After stepping out of the shower, I carefully extended my arm to grab for the towel draped over the shower/tub curtain. I walked around the lizard out to the bedroom, where, after drying off, I sat cross-legged on the floor at the edge of the worn bedroom carpet that met the bathroom linoleum. The lizard, apart from turning her head, did not move. She looked from my eyes to other parts of me, then back at my face. I had thought for sure that as I passed around where she sat on the rug she would run. She remained quietly watching. As I sat down, I had expected her to move. She didn't. My delight became shaded with worry. Maybe something was wrong with her. Or maybe she had been in the house all night and just didn't know the way out.

Finally I stood up and dressed myself, temporarily out of her field of vision as I took clothing from the closet. I returned to find her in the same position, the floor vibrations of my walking around and other movements evidently not making much of an impression. I wondered again if something were wrong. I softly approached her, gently extending my hand toward the rug. Slowly, curling her body, very snakelike in movement, the lizard began to turn and crawl off the corner of the white rug, at which point she turned back around and faced me—just out of reach.

I carefully knelt down and balanced myself with my left hand as I leaned forward. Feeling as if my arm were moving in slow motion, I again extended my right hand until the tip of my forefinger almost touched the point of her snout. The lizard was not watching my hand. Head slightly tilted, her eye was staring directly into my own. Suddenly her long, thin tongue flicked out, sensed the side of my finger, and pulled back in. I pushed my forefinger close enough to actually touch her snout. She didn't move. I slipped my finger under her lower jaw, gently stroking the smoothly scaled skin under her throat, then moved my finger down beneath her white belly. When she didn't move, I uncurled two other fingers, nudging my three fingers under her. The lizard now stared blankly as if at nothing. She crawled slowly onto my hand, but didn't stop. She continued until she had crawled halfway up my arm. And there she sat, tongue periodically flicking onto the tender skin inside the crook of my elbow.

Intending to carry her outside, I began to get up from my kneeling position. As I began to rise, the lizard unexpectedly jumped back down to the floor and scurried into the corner of a narrow area between the wall and the base of the bathroom sink. I waited for almost half an hour, at which time she crept out. I went to open the front door, then returned. She walked, swaying snakelike along the bathroom floor, until she reached the bedroom carpet. Sensing the change in texture, the lizard stopped. I offered my hand, placing it directly in front of her. She crawled up the back of my hand. I positioned my other hand to quickly grab her in a neck clutch between my thumb and finger should she attempt to jump while we were in motion. But she rode on the back of my hand until we reached the front door. Stooping my body and lowering my hand to the porch cement, I felt her little claws, and then her tail sliding over the back of my hand like a worm, as she crawled down. Finally the tip of her tail was no longer in skin contact. I lifted my hand and stood up. While not seeming to be in a hurry, the alligator lizard did walk at a steady pace until she turned the corner leading to the side of the manor, and I could no longer see her.

I stepped out and went to peek to see where she was going. But the lizard had vanished somewhere in the clumps of tall grass or between the cracks of broken cement or under a rock. I mused upon how remarkable it was to find a full-grown alligator lizard in the bathroom. She had probably crawled under the door to get into the house. She was the first alligator lizard that I had ever seen inside Petunia Manor. But she was not the last. However, I didn't know that at the time I set her out to go about her business, and I felt somewhat sad that Terry didn't get to see her.

To cheer myself, I lay down on an old, blue, vinyl lawn chair and watched the three jays chasing a mockingbird who had included, of all things, scrub jay calls in his immediate repertoire. Mockingbirds are endowed with the instinctive territorial ploy of being able to mock the songs and calls of other birds to imply sound-wise that the territory was already occupied by a given species. The jays would dive at and badger the mockingbird wherever he was, redwood tree or phone line or house top, until he would abruptly stop calling and fly. Then the jays would act as his escorts, swooping on either side and behind, until he landed. The jays would also land nearby. There would be a moment or two when all four birds, roughly the same size, sat quietly. Suddenly, the mockingbird, smallish gray head with beak pointed forward and upward, would begin his vocalizations, which set off the jays, and the whole game was repeated. I could not make up my mind whether the jays were chasing the mockingbird because he was on their territory or because they were not fooled but irritated by his scrub jay imitations.

I didn't come to a decision at that time, for a quick, sharp, localized—*and familiar*—nip of pain on my right buttock distracted me from the luxury of forming assumptions. I jumped up just in time to see the small, oblong, brownish black rear end with curved pinchers (what science books call *abdominal forceps,* as compared to *pincers*—which refers to appendages connected to the mandibles on the front of some creatures) of a male earwig disappear into and through the vinyl weave of the lawn chair. The little pain was gone almost as soon as it had come. There had been other times when I had looked to see the quick disappearance of a small, triangular head, mandible part facing me and pulling through the weave as the earwig prepared to drop, bottom end first, to the ground. Those little tricksters got me on the rear, always when I was relaxing face up in the lawn chair, a total of six times during those summers at Petunia Manor. The small, slim, shiny insects never bothered me when I lay stomach down, nor during the rare times before the morning when we realized that one had spent the night under the covers of the floor mattress with us.

Usually, especially in the summer, it was the occasional male earwig whom I found in the lawn chair—and who evidently must have felt the weight of my prone bottom to be a threat to his well-being. During this season, the female earwigs were typically not out and about. With their strong maternal instincts, the grown, female earwigs were usually under old wood items for sometimes two months tending and protecting young in various stages from eggs to nymphs who weren't ready to fend for themselves. The males, on the other hand, were foraging about for mites, various insect eggs, aphids, and rotting vegetable and fruit matter.

Human skin is so foreign from the earwig's natural diet and habitat that for the earwig to find any part of its body in contact intimately with any part of a human body, including the human ear, would probably be as instinctively stressful for the earwig as it is for the human. Whatever I felt those times on my rear must have been done by an earwig out of desperation. However, to this day, I do not know if the sharp, little pinches were the result of "pinchers" on the rear of the earwig's abdomen or the result of bites from the earwig. I picked up and shook the lawn chair to empty to the ground any earwigs who might be left within the vinyl weave before returning the chair to the porch.

After folding the lawn chair, I went back into the house and stared at the white rug in the bathroom. I knew that I would never have the unexpected joy of stepping out of the shower to find another similar, sweet, scaled visitor, as incidents like that are one-time gifts and usually caused by an error of judgment on the part of the creature. What I found that evening was almost more precious.

Engaged in the process of getting dinner, I was about to open the door of the refrigerator. A movement near the bottom of the stove nearby caught my eye, and I looked just in time to see a tiny tail tip slip around the side of the stove. Moving slowly and quietly, I took a few steps and leaned so that I could see around that side of the pale yellow stove. Creeping along the kitchen floor, as if determined to reach the living room carpet before I saw him, was the youngest baby alligator lizard that I had ever seen. He was probably no older than a month.

His head, like most baby reptiles, was proportionately large in relation to his tiny, slender body. The little lizard was so small that his checkered pattern was barely visible so that he appeared uniformly grayish tan. As I watched, he came to a stop, slowly lifting his head up and turning it sideways, staring back at me with one of his little, reddish, cat eyes. I noticed that in the middle of the top of his head was a dark cluster of scales that gave the appearance of a black blotch. As I was studying this unique marking, the young lizard suddenly burst into a blur of speed, moving fishlike off the kitchen linoleum and across the living room carpet until the tip of his tail disappeared under the couch. Knowing that he could easily get under the door to the outside when he was ready, I let him be.

A few evenings later, Terry walked into the kitchen to find, upon switching on the light, the same baby alligator lizard slinking across the kitchen floor toward a small, brown, rectangular rug at the end of the room. He captured the lizard and gently placed it outside on the porch. We didn't see the lizard for almost two weeks, at which time we again found the little thing scurrying across the kitchen linoleum. For fear of the baby lizard getting trapped under the stove, we put him outside again—this time, further away from the house.

But, to our astonishment, we found him yet again one evening in the kitchen. Three evenings had gone by since we had put him outside. The light switched on, he was already looking up with that black-dotted head, almost as if waiting, as if he knew that he was going to be put outside.

After awhile, as summer turned to fall, particularly by the end of October, finding the little alligator lizard crawling across the kitchen floor became more of an expectation than a surprise for us. He was now roughly a bit over three inches long, including his tail. His attempts to outrun us were less frantic. I like to think that this was because of familiarity, but I suspect that the colder nights affected the lizard's ability to move as fast as he did in the summer and early fall. By early November, because of the cold evenings, I had stopped putting the youngster outside when he was caught in the house after five o'clock in the evening. Instead, the lizard was placed inside a small trash basket, the bottom of which we had lined with dirt, twigs, and part of a shoebox so that he could sleep in privacy. Before leaving for the university in the morning, we would put him outside—usually at the back of the lot.

But the small alligator lizard always returned to the kitchen.

Near the end of December, we assumed that the tiny beast had finally gone into hibernation. A few weeks had passed since we last found him in the house. One day in early January, involved in the task of cleaning the house and about to sweep the floor, I picked up the small, brown rug at the end of the kitchen. I was startled to find the little alligator lizard, who was just as startled in his sleepy, slow-motion sort of way, uncurling himself. Terry and I were right. The young lizard had finally gone into hibernation. As he began to walk, quickening with every sneaky step, I gently replaced the rug. The small, moving lump beneath the rug slowed down until it was motionless. Then the shape changed. Because the lump was near the corner of the rug, I waited a few minutes then carefully and slowly picked up the rug edge. The little scaled body was curled like a snake's into a ball, pointed snout facing out from the cavelike opening made by my raising the corner. I gently let down the corner of the rug, then rummaged through the silverware draw and found a bottle cap. This I placed under the edge so that the rug was slightly raised enough to let air underneath.

So the alligator lizard, whom we called Magic because he was one of the creatures who exemplified the inexplicable, magical nature of those who dwelled with us at Petunia Manor, became part of our family for the next two months. We had to learn to walk softly when we were in the kitchen. The vibrations from our footsteps would often, though briefly, awaken him. Occasionally, I would see a change in the shape of the lump or, less often, the location. Sometimes I noticed the tiny, tan, pointed tip of his snout peeking out from the rug's edge or, on rare occasions, his entire head groggily protruding from beneath the rug. During these times, I gently offered water on a small spoon. Sometimes, Magic would slowly turn away, his head disappearing from view, and recurl himself under the rug. Other times, his long, thin tongue would almost aimlessly find its way into the spoon, briefly lap up water, then the tongue, the head with its unique, black mark in the center, and any other visible portion of the lizard would disappear under the rug. Usually, the lizard would settle down into a motionless lump beneath the rug at that point. However, as the days proceeded into March, after accepting a drink, Magic would retreat, but often formed an elongated moving lump under the rug for a minute or two before settling down.

Then one very early morning, Terry switched on the kitchen light, and there was Magic halted mid-step, one foot about to touch the floor. The little alligator lizard, standing unsheltered by the rug, was alert with his head tilted to the side, eye staring up at us. After this momentary pause, he casually began crawling across the kitchen floor toward the living room carpet—as if he were looking for a place to cross to the outside. I remembered when Magic had first come last August, evidently (but unknown to us at the time) to seek within the manor a safe, protected place for his very first winter sleep. The end of the lizard's hibernation brought about within me a dreamlike quality to that morning, and a feeling not unfamiliar, as this same feeling had been evoked by many of those creatures who dwelled with us within the shelter of Petunia Manor.

The Ravens

Ravens have no concept of unemployment. They are always employed at something—particularly play. We observed this in the forests near Petunia Manor and confirmed this fact years later when we lived in Prescott, Arizona. There are written accounts of observations of ravens using their bodies, feet tightly tucked to their sides, as sleds to slide down snow banks. While I have never witnessed this, I had anticipated doing so one day after a rare heavy snow in Prescott. The slope with the snow pack was a manmade one near a local shopping center. The high desert hill had been stripped of foliage, a large chunk gouged out to accommodate a parking lot, such that an edge of raw, red earth formed an artificial cliff. The low, steep cliff, which faced the large trash bins behind a few fast food restaurants, had become a social area for ravens.

The day I went to see if the account of ravens sledding could possibly be true, the area looked like a broad, white slide dotted with ravens. However, I did not see ravens sledding. What I saw was even more startling. The ravens were flying to the uppermost part of the sloped area where, feet tucked under and heads stretched back with beaks pointed upward, they rolled their bodies sideways down the slope. Some rolled over once and walked back up to the top. Other ravens rolled, blurs of black feathers, bodies turning over and over and over until the final roll when, mid-roll and still on their backs, the ravens would suddenly thrust their feet into the air, which allowed them to come to halt and remain upright, sometimes nearly at the edge of the parking lot. They would casually look around, heads tilted to one side and then the other, as if gaining composure, then fly back up to repeat the act. They reminded me of children, bodies prone, rolling down grassy hills. There was no purpose in these ravens but play.

Along with their capacity for play, ravens have a tremendous sense of humor. And this is why, during our last year at Petunia Manor—schooling completed, respective degrees in hand, unemployed, and the future uncertain—the ravens kept for us the sense of humor we were losing.

The prospect of finding *real* jobs, moving, and, in general, the feeling of chaos that accompanies the end of the academic routine, forced Terry and me to often leave the abstractions of our fears behind by walking among the tall redwood trees that dwelled in Armstrong Woods to re-establish reality. Feet literally on the ground, we had found stability in our paths in the form of a family of ravens. The two adults and three young were unchangingly present in the grove, usually walking busily about the picnic tables and beaking with intention any items appealing to their interests, including one another. The wispy, muted clapping of flapping wings often accompanied the soft crunching of our own steps.

We usually heard the ravens before we saw them, and their splendid noise determined the direction of our walk on a given day. Caws, clicks, croaks, screams, squeals, snaps, grunts, squawks, barks, purrs, and an assortment of indescribable, miscellaneous sounds filled the air between the picnic tables and the tops of the trees. I once heard it said that, next to humans, ravens have the largest number of spoken dialects. I speculated that one of the variants of raven dialects, in addition to vocalizations, is degree of loudness.

At times, the tops of the trees would ring with the cacophony of loud, nonstop vocalizations. The amazing thing was that most tourists to Armstrong Woods seemed not to notice. People, oblivious to the auditory riot taking place above them, would blissfully take one another's pictures under the largest redwood trees.

So, when one day we came to the redwood grove and it was quiet, we began to suspect that the ravens had flown to new territory. It was late in the summer, and this seemed like something we would have to do as well, for their were no jobs in our field to be found in Sonoma County. The seeming disappearance of the ravens, who had been our daily diversion from dwelling too much on the more practical matters of life, seemed to magnify with intensity our own desperation. Nevertheless, we still decided to walk the loop paths of the grove, our feet moving us in circles, and our talk following in kind, with long periods of silence.

We had been walking quietly, holding hands. We momentarily stopped under a canopy of redwood branches. I let go of Terry's hand to pull up one of my socks. The moment we unlocked hands, a huge piece of branch fell from the tree above, dropping squarely down the thin space between us. We both looked up, only to find what one would expect to find—branches of trees hiding smaller branches that led to the tops of trees, which could not be seen from the ground where we stood beneath. As I once again bent down to pull up my sock, a large but light flake of bark dropped down, bouncing off my back and onto the ground. Terry and I looked at each other in surprise, then fixed our gaze once again on the upper branches of Redwoods directly above the path.

The first falling of something from a tree can be considered coincidence. But, if it happens immediately again, particularly when it is a direct hit, this may safely be considered with nothing less than suspicion. But, once again under our scrutiny, the tangle of branches above were free of anything suspicious, innocently still.

Terry walked a few steps to the side of the path to sit down and wait on a redwood stump as I continued to fidget with my other sock. He had not been sitting there for more than a few seconds when a flurry of bark bits and small sticks floated down from the end of one of the branches to catch in his hair. I watched in amazement.

He was already standing and looking up again. I joined him, thinking that somewhere up there a squirrel was scurrying about. But the somewhat suppressed "laughing" sound that came from another branch above caused that thought to leap from my mind. While the sound resembled human laughter, in an auditory caricature sort of way, I suddenly realized that it was one of those unclassifiable, miscellaneous sounds so typical of ravens.

I looked up again, this time with expectations. But in my close surveyance of the dense canopy of branches, my eyes found no ravens. But I did catch a movement. And with the movement came a suspicious sound, almost like a whispered squawk. I walked up the path to get a different visual perspective. And on a branch, which had been hidden from view before by other branches, lined up close together like a string of black feathered balls with tails and all looking down at us, were four untypically quiet ravens. We could hear the fifth raven flapping on a separate branch nearby in the shadows, but we could not see him—until he, making barely audible guttural noises, proceeded to hop to the branch with the other ravens, where he positioned himself directly above us and let go a beakful of dried moss and bark bits. As the shower landed on us, evidently the silence code was broken, for there was a burst of high-pitched caws, barks, clicks, and shrill squawks, sounds perfected only by ravens, then the familiar swoosh of wings flapping.

One raven, a molting adult, remained, stepping sideways up the branch and tilting her head comically to stare at us. A small wisp of grayish black feather fell from her neck as she continued, big beak now closed and silent, to look down at us with that gleaming bead of an eye. Anyone who has the experience of seeing eye to eye with a raven for any extended period of time comes to know that life cannot possibly be as serious as humans want to make it. Ravens must find our attempts to do so the very thing that makes us so amusing to them, for why else would they make us, at times, the targets of their play?

Indeed humans are too serious for this world. They are self-appointed victims of a mind set that play is inferior to work. Humans also have a tendency to be deadly serious about territories and possessions—and, as in the case of some new residents to the Mojave Desert, the contents of their backyard garbage containers.

In 1990, there was a grave misunderstanding. The problem occurred because the humans did not understand that one creature's trash is another creature's toy box, and the ravens did not understand that people's trash came with boundaries.

We had heard about the problem through the Ecology Club when we were still at the university. Evidently, many people who had moved from Los Angeles to the newly built residential areas in the middle of the desert were indignant because they could not guard successfully their discarded possessions, trash as it was, from foraging ravens. The city commuters had moved to the desert because homes were cheaper out in the middle of nowhere. But nowhere was somewhere to the ravens who had been dwelling there for generations before Los Angeles was created. The birds seemed almost gracious in their accepting, in return for the loss of their territory, the new perches of telephone poles but, even more so, the food and play things put out in odd containers near the end of the yards attached to the boxes in which the humans lived. And, to the ravens, the areas that contained these treasures were somewhere to be.

The problem was humorous, but the rumored solution was hideous. Because a certain bureau of the government was getting pressured by whining residents who were distressed at the thought of garbage sharing, the bureau had to propose something. So, it was rumored, someone invented an ingenious reason to rationalize the proposal to randomly place arsenic in hard boiled eggs and leave them about during raven nesting season. This solution would effectively eliminate both the ravens and their young—but leave the trash in the garbage containers intact.

The plan consisted of starting a war between ornithologists and herpetologists, and at the same time, promoting to the public the raven as a killer of the endangered desert tortoise. Because I had volunteered for over a year in a vivarium which contained a couple of desert tortoises, I new that, unless the tortoise was very young, very old, or very sick, there was no way that a raven beak, big as it is, could penetrate that hard calloused shell once the tortoise had withdrawn. Now, the fact that tortoise killings by ravens were supposed to be happening on a mass scale in the Mojave Desert made me somewhat cynical of the motives of those people who perpetuated the stories. Many of us at the university thought it odd that this alleged problem had not been noticed prior to the ever increasing housing developments in the desert.

It happened to be that, at the time the residents-versus-ravens-over-garbage talk was floating around campus, some of us were taking different sections of a Comparative Religion course in which one of the projects was participating in a religious experience of one's choice. Some students explored Catholicism, others explored Judaism, and still others visited a Buddhist temple. A group of us, using the excuse that we were exploring paganism, headed up the coast to the Fort Ross area. There, Terry and I shared with our companions a place about a half mile walk into the conifers where a government agency dumped, for some reason unknown to us, massive quantities of fish remains. The significance of this area in the forest was that, at any given time, one could be in the middle of ravens and vultures who, attracted to the fish remains, hung out in large numbers.

Our little group sat in a circle, joined hands, and, to the backdrop of the sound of shifting and flapping raven wings in the trees above us, sent our thoughts of warning to the ravens above. The idea we fancied was that these ravens would somehow carry the warning of the arsenic-heavy scheme to their relatives in the desert. Fanciful or not, it was the only thing we felt we could do from a distance.

The poisoning of the ravens, in fact, never came to be—the controversy, no doubt, being too risky to the reputation of those who would dare to carry out such a wretched deed. The incident was soon forgotten, as was our little pseudo-ritual.

I hadn't realized until the day that the ravens were playing with Terry and me in Armstrong Woods two years later that ritual and intention are nothing to take lightly. For, as near as I can tell, the day we joined in the warning ritual as their friends was the point in time that we developed our special bond with the beautiful, black masters of play—the ravens.

The Last Spider Roundup

I have been a herder of many creatures: cattle, horses, sheep, goats, and chickens. But of all the two-legged and four-legged critters I tried to coax from one point to another, never was there an odder bunch of herdees than those to be transported from Petunia Manor to safer quarters.

We're talking arachnid roundup, here. Get along little spiders...

The prospect of such a drive of long-legs dampened our sudden sense of prosperity which seemed imminent. There was, in that fourth year in Petunia Manor, with college degrees safely in tow, a real likelihood of livelihood on the near horizon. This caused our hearts to swell with hope. Until reality woke us up to the larger implications of abandoning Petunia Manor.

Over the years in that loveable hovel it had been transformed by our affection for all things living from a lifeless, bug-bombed out shed to a haven for spiders of many species. Or what we had come to refer to as many nations. Each room was busy with elaborate moldings made of gossamer gauze—the sheet webs of generations of cellar spiders who had spun, bounced, courted, dined, molted, raised constellations of babies, and, sometimes, died. These spiders had taught us many lessons, among them, the truth of Thoreau's assertion about economy: wherever possible, address your creaturely needs by careful use of native materials. The spiders certainly practiced this, creating homes from their own essence, repairing and restoring existing webs, and finding sustenance from local game—flies, ants, mosquitos, and moths. The spiders also practiced the difficult art of patience. We watched them during lean times, such as winters when insects were virtually absent, as their elegant bodies shriveled as they hung in a kind of suspended animation in their quiet webs. This was a beautiful example to us of conservation of energy.

It was one of the spiders' many teachings. During periods of deprivation, pull in, become quiet, contemplative. Play upon the back of your eyelids (all two or four or six or eight) your dreams of simple things and sustain yourself on the meaty vapors of faith. Almost always, it seems, nourishment comes along just about the time the walls of your abdomen are about to meet. This was a lesson we took to heart during that last, long, fretful year at Petunia Manor. We had labored for three years to complete our academic preparation for our foray into education. But as the months ticked by and graduation appeared on the nearer horizon so did the knowledge of the even bigger challenge ahead. This was a terrible time to be trying to break into the teaching profession, we were told. There was something akin to a taxpayer revolt. Funding for education was down. In California there were massive lay-offs, department closings. My early applications for the few openings confirmed this. There were often from 200 to 400 applicants per slot.

As if that weren't enough, our battle against developers who wanted to transform the Laguna uplands into pricy real estate had taken a turn for the worse. It seemed that the proposal would be approved and construction (destruction) would begin within months. Then we got a notice from our landlord. It was a friendly missive, but firm. Due to the rapid acceleration of his property values, due to the impending development, he would have to raise the rent. Raise it beyond our ability to pay. It had now become a race to find gainful employment before we would be evicted.

"Remember the spiders," Eve said softly. I tried. But somehow I couldn't quite summon the inner-strength of my eight-legged counterparts. Faith, it seemed, was not a sustaining meal.

We consoled ourselves by taking more walks along the imperiled Laguna. But this served to burden my heart even more. I watched the green, green grasses alive with life and shook my head and lamented that I couldn't even succeed in saving these splendid beings from the destruction that seemed now sure to be visited upon them. Finally, we made our way to the grove of Redwood Elders that we had discovered those ten long years before upon first setting feet in the region. It was there, in the hushed cathedral of living antiquity, that voices reached out for us from above and, in that inimitable raven way, made us know that it was gonna be all right.

A few days later, the phone rang. It was an old friend who lived in a small town in central Arizona. There was a job opening, he said. It's not exactly what you're looking for, but you might have a good shot at it.

The story of that frenzied journey to Prescott, Arizona and related events is itself a small miracle. But that will have to be told elsewhere. Let it be said for now that by the time we returned to Petunia Manor three weeks later (a hurried trip up to Wyoming was sand-wiched between rounds of interviews), we had joined the ranks of the newly employed. The part-time position had grown up into a full-time one. Eve's talents won her a position at a neighboring college. And we were expected back for duty within ten days. Our joy raised the ceiling of the tule fog by a few feet.

And then it struck us. Hard. Packing up all of our worldly belongings could be easily accomplished with long days and Ben Gay nights. But once we pulled out of the narrow dirt lane and bid Petunia Manor goodbye forever, we knew that the landlord, God bless 'im, would be ready to pounce with bug bombs and high-pressure washers, and who knew what else. All of our remaining friends of eight legs, four legs, two legs, and no legs would be directly in harm's way. The very creatures who had sustained us through the long dark days of our own liberation would themselves face eviction at best, destruction at worst.

Almost simultaneously, we formed an idea of simply moving the spiders from the manor. Our flame of hope was rekindled. Until I did a quick once-over inside and out and tallied a resident Petunia Manor spider population of perhaps a thousand! How on earth could we possibly complete the packing, loading, house and yard cleaning, and spider relocation—all within a mere week! We simply couldn't answer that from a rational, reasonable stance. We would have to do as the spiders had taught us: in terms of the challenge, make it small. Simply begin with a heart filled with hope. And never look back.

And so it began.

The first big question was where? Where can we relocate the spiders to? We called upon our friend up the street who had become a staunch ally in the fight to save the Laguna. He already had in his house a bountiful bestiary of native creatures and domestic animals. Would he consider providing sanctuary for a few (ahem) more eight-leggers?

Yes he would. Bless him. While Eve remained inside the manor to fashion a spider shuttle, I toured the outside of the beige little house and took a more accurate count of the relocatees. The scope of the mission became truly daunting as the hundred or so on each of the east, south, and west sides were now joined by something like three or four hundred spiders on the north face of the manor.

There were adults and teenagers and brand new, delicate babies whose new little legs were still clear and smaller than an eyelash. The weight of responsibility made me bow-legged, squashed my spirit flat. Atlas, by comparison, was holding up a Nerf ball in comparison to the weight now squarely on our shoulders. The most gripping reality of all was this: if we are to save one, we must save them all. There could be no convenient appeal to compromise here. Each and every one of these spiders was a true innocent. They had no idea of what lay in wait for them. So just as they had helped us to heal from our many wounds and heartaches over the years here, we must lend the helping hand to each of them. And we had thought that getting a job was the big challenge!

Thanks to Creation for our human friends. They proved their love for us by turning out to help us pack, stack, and clean. And they offered this support in full knowledge of what I was busying myself with during the whole moving operation. There was some gentle laughter, even a rolled eye or two. But through it all our friends showed that they were truly friends—they love you no matter HOW STRANGE your actions seem.

So, over a long weekend beneath the hazy glare of an August sun, while friends stashed our books, plates, shoes, kitchen utensils, and bathroom sundries in cardboard boxes, Eve and I completed the design for the great spider relocation campaign.

The prototype of the spider transport vehicle had already undergone extensive test drives. It was a Saffola margarine tub.

Eve had long ago perfected a method of capture and release that involved slowly moving the plastic container up alongside of the spider, one side of the tub being leveled against the wall to offer a convenient "step on" ramp for the passenger. She would pause the tub right at the spider's feet, letting the spider feel the plastic and hopefully adjudge it non-threatening. After a moment, the lid would be slowly moved up to the spider's rear flank, pausing, then gently touching the spider's rear foot. Most times the spider would walk right into the shuttle. Usually, the lid did not need to be administered. The spider would be inspecting the old traces of webbing that clung to the white plastic. Spiders, we had felt, read those webs like newspapers. The bottoms of spider feet are sensitive, and somehow the spiders get all kinds of information from the webs. If the spider seemed upset, the lid would be fitted loosely. This latter step is recommended for folks who aren't yet accustomed to the tickle of spider legs walking over their own skin, if the long-legged one slips from the shuttle and makes an escape over the rough landscape of the human hand and arm.

On this moving day, the sheer numbers of relocatees demanded a different design. We commandeered the resources of packing boxes, realizing that the broader "on-ramps" offered by the cardboard sides would better serve the loading process. Lids would not be necessary as the vast interior of the box would offer plenty of exploration for the spiders while they were being walked up the street to their new home.

The process was simple. For our inside spiders who were more familiar with us, we created a small ritual. Eve proceeded me to the corner where several or a dozen spiders slumbered in their sky beds. She would gently blow on the silky home. This is the equivalent of a doorbell in spider land. The spiders would wake up and often go into immediate bounce mode. Once the bouncing subsided and the spider(s) became visible again, she would offer up an explanation of what we were about to do, followed by an apology and topped off by a small prayer of thanks for their presence in our lives, and our fervent hopes that their lives would be good in their new homes. Nearly all of the long-legs seemed to understand what we were doing. Most were very cooperative and with only a bit of coaxing stepped onto the brown cardboard escalator. I would get a half dozen or so in the box shuttle and then take off down the alley, bound for the street and, eventually, for our friend's waiting sanctuary.

He was waiting for us with a kind of bemused grin, holding the door open wide to admit the moving "van" and its stirring passengers.

"So how many more you got?"

"Uh—well, a few. You got some outside spaces, too?" I asked hopefully. Our friend indicated that his old, tight-board fence and the hedge of trees out back afforded many a spider shelter. I secretly calculated that we might find homes for perhaps a couple of hundred spiders here, but I kept the number to myself for now.

The cardboard moving van made perhaps four dozen more trips, with as many as fifty spiders inside. Most of the occupants rode in relative calm. But the occasional long-legger, not lulled by the jerky rhythm of the beast of burden, made a break for it, boiling up over the rim of the box and out onto the outer surface. There it would rake the air with an uncertain leg, trying to make sense of this involuntary odyssey. One actually bailed off in the middle of the asphalt avenue and I threw down a hanky, waited for the escapee to run on to it, then drew the four corners of the hanky together and forced the enthusiast to finish the trip in this cotton cradle.

On one of the trips I thought of how empty the box felt, yet how jammed with life it was. It reminded me of those times inside Petunia Manor when we relocated a spider because it was in harm's way—in the shower stall or trapped by the stainless steel cliffs of the kitchen sink. Often we would simply let the long-leg hitch a ride on an outstretched finger or palm. The spider would hesitate at first, apparently not enjoying the taste of human skin cells. But usually spiders would hop aboard, often scampering over our hand, up an arm, barely tickling the individual hairs with their featherweight bodies. It was then we would marvel at the great effect these creatures had on our lives for being nearly insubstantial in body. It was another small lesson: one can throw one's weight around to garner attention, or one can simply act on the true faith of life properly lived and depend upon one's good effects to leave a memorable impression behind. Cast in terms of beneficent effect, these spiders were as big as elephants. With their bounces and courtship dances and elegant weavings, the cellar spiders of Petunia Manor had beat the laws of physics and, on many an occasion, helped us to dissipate the gravity of our heavy moments. This act of relocation somehow seemed like a rightful reciprocity. The spiders had lifted our spirits on numerous occasions. We now returned the favor by lifting their bodies to safe haven.

The clans of Petunia Manor's inside spiders were safely relocated by mid-morning. I now turned my attention to the outside of the little house. There were far too many thereupon to liberate within our friend's house. These spiders would go from exterior to exterior domiciles. For three hours I gently herded spiders into the box and carried them off to release sites on wooden fences, mailbox posts, along a sagging shed down on the shoulder of the Laguna. Before the relocation was finished, there was scarcely a vertical surface in the neighborhood that had not been given the gift of a Petunia Manor spider. Over the years we had kept track of which neighbors did not use toxic chemicals in their yards. We figured in return for their care of the land, they deserved some free spiders to help them with insect control. Good gardeners know that spiders are one of nature's most beneficial creatures, as they regularly dine on aphids and other sap suckers.

So absorbed was I with the relocation that I forgot completely about the human world around me. At one delivery site—a stand of small redwood trees along the avenue about a half block away—I happened to look up from my work to see a neighbor lady watching me intently from behind her screen door. I smiled and waved with my free herding hand, which, I realized, bore a big female cellar spider. This caused me to withdraw my hand and speed it to the trunk of the tree. I set down the box and used my other hand to coax the spider off my fingers and onto her new home. From the woman's vantage point, I realized, this must have looked like a friendly wave turned suddenly into some kind of weird and secret communion with the bark of the tree. Tree hugger? Tree stroker?! I smiled again and saw the woman duck back into the shadows of her house. I further imagined what she had been witnessing: a man is seen gingerly carrying a cardboard box. He stops at a tree and carefully opens the box. Then, it's like he thinks he's harvesting some kind of fruit and putting it into his box. I was lucky she did not summon the authorities.

There is a way in which this harvest fantasy was true. By relocating the cellar spiders to places all over the neighborhood strung along the Laguna, there was for us a harvest of hope. Hope that the spiders would thrive in their new nooks and crannies. Hope that the magic they had spun for us in Petunia Manor would now be woven for the entire neighborhood. And that this magic, along with that of the Three Sisters, would extend beyond the little houses and yards to the Laguna uplands to spare that precious stretch of land the destructive enactment of a developer's scheme.

The big moving truck was loaded to the gills with our worldly possessions. The little shed that had served us so well as a home was now washed and swept—a sparkling, empty shell echoing as a beach shell. Friends were thanked and hugged. We spent a restless night rolled up in our sleeping bags on the floor of the bedroom where we had received so many dreams, been blessed with the reflected, living light of the Redwood outside the east window. Morning came, bright and blue. Now, on the brink of our departure for far off Arizona, there remained but one final act to perform in Petunia Manor.

We searched down the avenue among the cinnamon scimitars of eucalyptus leaves until we found what we sought: a perfect, blue scrub jay feather. It wore the exquisite, deep blue of the new day's sky. Our footfalls echoed eerily in the empty rooms. We stood, side by side, in the empty bathroom. Facing west, we turned our eyes upwards to the southwest corner where ceiling kissed walls. There, still largely intact, were the shimmering folds of Goldman and Webman's sky farm. There was the silken cradle upon whose etheric brow still rested the tiny molts of the last generation of babies. They shone in the morning light like fragile golden flowers.

Every other web had already been gently removed from the house and draped over a nearby fence or tree branch or flower blossom. Only this web remained. For in this web had lived the two ladies who weaved so much magic into our lives, who had brought solace and wisdom to our frantic days, our desperate or despairing moments.

We held hands and said a few, simple words of gratitude to the memory of the two grand ladies of long legs and infinite grace. We offered up hopes for the well-being of all their children and children's children. As long as the descendants of Goldman and Webman could spin their magic here, we believed, all would be well with the Laguna and this little community on its grassy shoulders, and in and around the squat shed we had come to call Petunia Manor.

Eve touched the scrub jay feather to the web and gently rotated it, until, like a spindle, it gathered to itself the web. The last strands pulled away from the walls and then the entire web, the cottony castle of the Spider Queens, lay securely wrapped around the blue feather like cirrus clouds draped on a turquoise sky.

I followed Eve out of the house, shut the door softly behind us. We proceeded to the base of the redwood tree. At its base, half nestled in a deep crevice in the furry red bark, Eve lay the feather.

We crawled into the cab of the moving truck. I fired the engine. We drove slowly down the narrow, graveled lane to whatever grand future that lay ahead.

A postscript:

Once in Arizona, safely ensconced in a house, we were continuing to unpack cardboard boxes one afternoon. I opened one to find a tangle of kitchen utensils. As I extracted knives and forks and ladles, I detected movement across one flap of the box lid. To my utter astonishment, out walked a mature, long-legged cellar spider lady! She moved deliberately down the outside of the box, across the kitchen counter, up the wall, until she came to the alcove formed by the uppermost region of the kitchen's southwest corner. There she began to web. To create her new home. I summoned Eve and we watched her weave with confidence and poise. We knew, then, that this new life in Arizona was to be as filled with magic as were those long and memorable days in Petunia Manor.

THE END